Famous Volcanoes

Unveiling the Power and Beauty of Earth's Fiery Giants

© 2023 Jordan Y. Wagner.

All rights reserved. No part of this book may be reproduced, stored in a retrieval system, or transmitted in any form or by any means, electronic, mechanical, photocopying, recording, or otherwise, without the prior written permission of the copyright holder.

This book, "Famous Volcanoes: Unveiling the Power and Beauty of Earth's Fiery Giants," was created with the assistance of Artificial Intelligence. The information provided in this book is for entertainment purposes only. While every effort has been made to ensure the accuracy and reliability of the content, the author and publisher make no representations or warranties of any kind, express or implied, about the completeness, accuracy, reliability, suitability, or availability with respect to the book or the information, products, services, or related graphics contained in the book for any purpose.

The author and publisher disclaim any liability for any loss or damage arising from the use of this book or reliance on the information contained herein. Any reliance placed on the information in this book is strictly at the reader's own risk.

All pictures used in this content are sourced from Canva and are considered license-free. Canva provides a platform that offers a variety of images that can be used for personal and commercial purposes without infringing on any copyright restrictions. The use of Canva images in this context is in compliance with Canva's terms of service and licensing agreements.

A Fiery Introduction: Unveiling the Mysteries of Volcanos 6

Titans of Fire: The Majestic Mount Vesuvius 8

Roaring Beauty: Exploring the Power of Mount Fuji 11

Nature's Fury: The Cataclysmic Eruption of Krakatoa 14

The Peak of Mysticism: Mount Kilimanjaro's Eternal Charm 17

The Sleeping Giant Awakens: Mount St. Helens' Explosive Rebirth 20

The Pillar of Smoke: Discovering Mount Etna's Enigmatic Nature 24

Flames on Ice: The Intriguing Saga of Eyjafjallajökull 27

The Island Builder: Mauna Loa's Mighty Lava Flows 30

The Sacred Fire Mountain: Pele's Realm of Kīlauea 33

A Tale of Destruction: Mount Pinatubo's Devastating Eruption 36

Beyond the Clouds: Reaching the Summit of Mount Cotopaxi 39

The Gateway to Hell: Mount Nyiragongo's Incandescent Crater 42

The Jewel of the Caribbean: Soufrière Hills and the Destruction of Montserrat 45

A Charming Fury: Mount Agung's Volcanic Ballet 48

The Frozen Inferno: Mount Erebus and Antarctica's Fiery Secret 51

The Sleeping Giant of Europe: Mount Teide's Majestic Silence 54

The Great Pacific Fireworks: Exploring Popocatépetl's Explosive Show 57

In the Shadows of Giants: Hekla's Mysterious and Frequent Eruptions 60

The Ring of Fire's Wrath: Unraveling the Tale of Mount Rainier 63

The Island's Sentinel: Mount Yasur's Mesmerizing Lava Display 66

The Forgotten Volcano: Mount Tambora's Eruption That Shook the World 69

The Phoenix's Nest: The Birth and Evolution of Santorini's Caldera 72

The Jewel of the Andes: Discovering the Beauty of Mount Chimborazo 75

The Firebird's Domain: Exploring the Intricacies of Mount Nyamuragira 78

Epilogue 81

A Fiery Introduction: Unveiling the Mysteries of Volcanos

Volcanos, the magnificent and awe-inspiring forces of nature, hold a captivating allure that has fascinated mankind for centuries. These remarkable geological formations are windows into the dynamic and restless nature of our planet, showcasing its immense power and creative potential. In this chapter, we embark on a journey to unravel the mysteries surrounding volcanos, exploring their formation, characteristics, and the awe-inspiring phenomena that occur within their fiery depths.

At the core of every volcano lies the Earth's molten mantle, a seething mass of rock that constantly churns and moves beneath the surface. Volcanos form when this molten rock, known as magma, rises towards the Earth's crust, propelled by the intense heat and pressure generated within the planet's interior. As magma ascends, it seeks out weak points in the Earth's crust, often found at tectonic plate boundaries or areas of volcanic activity. When the magma breaches the surface, it erupts, giving birth to a volcano.

The structure of a volcano is a complex interplay of various elements. The central vent, or the main opening through which magma and gases are expelled, is the focal point of volcanic activity. Surrounding the central vent are layers of hardened lava, ash, and other volcanic materials, forming the cone-shaped structure commonly associated with volcanos. Crater, a bowl-shaped depression at the summit, may be present, resulting from the collapse of the central vent or subsequent volcanic activity.

Volcanos exhibit a wide range of eruption styles, from gentle effusive eruptions to explosive cataclysms. Effusive eruptions occur when magma flows relatively smoothly from the volcano, allowing it to release gases and lava gradually. These eruptions often result in the formation of shield volcanos, characterized by their broad, gently sloping profiles. On the other hand, explosive eruptions are violent and spectacular events, involving the rapid release of trapped gases and fragmented magma. These eruptions can generate towering ash clouds, pyroclastic flows, and volcanic projectiles, transforming the landscape and endangering surrounding areas.

One of the most significant factors influencing volcanic activity is the composition of the magma. Magma is primarily composed of molten rock, but its exact composition can vary greatly depending on the concentration of different elements and compounds. The presence of silica, a mineral compound, is particularly crucial in determining the viscosity of the magma. Magma with high silica content tends to be more viscous, resulting in explosive eruptions, whereas magma with lower silica content flows more easily, leading to effusive eruptions.

The effects of volcanic eruptions extend far beyond the immediate vicinity of the volcano. Volcanic ash, a fine powder consisting of tiny rock fragments, can be carried by winds for great distances, blanketing vast areas and disrupting air travel. Volcanic gases, such as sulfur dioxide, carbon dioxide, and water vapor, can have far-reaching environmental and climatic consequences. Sulfur dioxide, for instance, can react with water vapor in the atmosphere, forming sulfuric acid, which contributes to the formation of acid rain.

Despite the potential dangers they pose, volcanos also play a crucial role in shaping the Earth's landscape and providing fertile grounds for life to thrive. Volcanic activity releases essential nutrients and minerals into the surrounding soil, enriching it and making it highly fertile. Volcanic islands, formed by the accumulation of volcanic materials over time, often boast lush and diverse ecosystems, showcasing nature's ability to adapt and flourish even in the most challenging environments.

Throughout history, volcanos have left an indelible mark on human civilization. The archaeological ruins of Pompeii and Herculaneum, preserved beneath the ashes of the catastrophic eruption of Mount Vesuvius in 79 AD, offer a poignant reminder of the destructive power of volcanos. Volcanic eruptions have shaped landscapes, influenced climate patterns, and even played a role in shaping cultural beliefs and mythologies across different civilizations.

As we delve deeper into the chapters ahead, we will embark on an extraordinary journey to explore the world's most famous volcanos. From the ancient ruins of Pompeii to the icy peaks of Mount Erebus in Antarctica, we will witness the raw power and beauty of these geological wonders. Prepare to be captivated by their stories, intrigued by their mysteries, and amazed by the sheer force of nature that shapes our planet.

Titans of Fire: The Majestic Mount Vesuvius

Perched majestically on the Gulf of Naples in southern Italy, Mount Vesuvius stands as an iconic symbol of both beauty and destruction. This legendary volcano has etched its name in history through its catastrophic eruption in 79 AD, which buried the ancient cities of Pompeii and Herculaneum under layers of ash and pumice. In this chapter, we delve into the captivating story of Mount Vesuvius, exploring its geological significance, historical impact, and the mesmerizing allure it holds for visitors from around the world.

Mount Vesuvius is part of the Campanian volcanic arc, a chain of volcanos that stretches along the western coast of Italy. It is classified as a stratovolcano, characterized by its steep slopes and explosive eruption style. Standing at an imposing height of approximately 4,203 feet (1,281 meters), it dominates the surrounding landscape and offers breathtaking panoramic views of the Bay of Naples, the bustling city of Naples, and the picturesque islands of Capri and Ischia.

The history of Mount Vesuvius is intertwined with the ancient Roman cities that thrived in its shadow. Pompeii, a bustling commercial hub, and Herculaneum, an elegant seaside resort, were both tragically entombed in ash and debris when Vesuvius erupted in 79 AD. The eruption, which spewed a column of volcanic ash and gases miles into the sky, remains one of the most infamous volcanic disasters in history. The suddenness and intensity of the eruption caught the inhabitants off guard, resulting in the preservation of their everyday lives for centuries to come.

Today, visiting the ruins of Pompeii and Herculaneum offers a glimpse into the past, where time stands still amidst the remnants of ancient Roman life. Walking through the well-preserved streets, homes, and public buildings, one can marvel at the vibrant frescoes, intricate mosaics, and the haunting plaster casts of the volcano's victims, forever frozen in their final moments. These archaeological sites provide invaluable insights into the daily routines, social structures, and artistic achievements of a bygone era.

For those seeking a closer encounter with the volcano itself, a journey to the summit of Mount Vesuvius is a must. Guided tours and hiking trails lead intrepid explorers through the rugged terrain, offering panoramic views of the crater and the surrounding landscape. The volcanic landscape, characterized by its barren slopes and hardened lava fields, stands as a testament to the volatile nature of Vesuvius. From the summit, visitors can peer into the crater, a gaping hole that serves as a constant reminder of the potential for future eruptions.

When planning a visit to Mount Vesuvius, it is essential to consider safety precautions. As an active volcano, Vesuvius is continuously monitored by scientists to detect any signs of impending activity. Visitors should check the local authorities' guidelines and restrictions to ensure a safe and enjoyable experience. It is advisable to wear appropriate footwear and clothing, as the terrain can be rugged and dusty. Carrying water and sunscreen is also recommended, as the sun can be intense on the exposed slopes.

To enhance the experience further, combining a visit to Mount Vesuvius with a tour of the nearby archaeological sites and the vibrant city of Naples is highly recommended. Naples, renowned for its rich history, art, and mouthwatering cuisine, offers a gateway to the region's cultural heritage. Exploring the National Archaeological Museum of Naples, which houses a vast collection of artifacts from Pompeii and Herculaneum, provides a deeper understanding of the region's ancient past.

In conclusion, Mount Vesuvius stands as a titan of fire, both captivating and haunting in its power. Its historical significance, geological marvels, and breathtaking views make it a must-visit destination for history enthusiasts, nature lovers, and adventure seekers alike. As you stand on its slopes or gaze into its crater, you cannot help but feel a sense of awe and reverence for the forces that shape our planet and the indomitable spirit of human resilience in the face of natural calamity.

Roaring Beauty: Exploring the Power of Mount Fuji

Rising gracefully from the pristine landscape of Japan, Mount Fuji stands as a symbol of beauty, spirituality, and national pride. With its iconic snow-capped cone piercing the sky, this majestic stratovolcano captivates the imagination of all who behold it. In this chapter, we embark on a journey to unravel the power and allure of Mount Fuji, exploring its geological significance, cultural importance, and the awe-inspiring experiences that await those who venture to its slopes.

Located about 100 kilometers southwest of Tokyo, Mount Fuji, known as "Fuji-san" in Japanese, is Japan's highest peak, towering at an impressive height of 3,776 meters (12,389 feet). Its near-perfect symmetrical cone shape has been shaped over thousands of years through volcanic activity. Despite its seemingly serene appearance, Mount Fuji is an active volcano, albeit in a dormant state, with its last eruption occurring in 1707 during the Edo period.

The geological importance of Mount Fuji lies in its formation as a stratovolcano, which is characterized by layers of hardened lava, ash, and volcanic debris. Its volcanic activity is associated with the subduction of the Pacific Plate beneath the Eurasian Plate, a tectonic process that has shaped much of Japan's geological landscape. Over time, the repeated eruptions and subsequent cooling of lava have built up the distinct cone shape we see today.

Mount Fuji's cultural significance cannot be overstated. It has been a source of inspiration for countless artists, poets, and writers throughout Japanese history, often depicted in traditional ukiyo-e woodblock prints and haiku poetry. Its beauty has been celebrated as a symbol of purity, strength, and spiritual enlightenment. The mountain holds deep religious importance in Shintoism and Buddhism, with pilgrimage routes leading to its summit, attracting devotees seeking personal introspection and enlightenment.

For those seeking to explore the magnificence of Mount Fuji, there are numerous ways to experience its grandeur. The climbing season typically runs from July to September when the weather conditions are most favorable. The most popular trail is the Yoshida Trail, starting from the fifth station and winding its way up to the summit. This route offers stunning views of the surrounding landscapes, including the Fuji Five Lakes, which provide picturesque reflections of the mountain on calm days.

It is important to note that climbing Mount Fuji requires adequate preparation and physical fitness. Hikers should come equipped with appropriate hiking gear, warm clothing, sturdy footwear, and enough food and water for the ascent. It is advisable to check weather conditions and consult local authorities before embarking on the climb. Additionally, there are mountain huts along the trails where climbers can rest and refuel during the ascent.

For those who prefer a more leisurely experience, Mount Fuji can also be appreciated from various vantage points in the surrounding area. The Fuji Five Lakes region, comprised of Lake Kawaguchi, Lake Yamanaka, Lake Saiko, Lake Shoji, and Lake Motosu, offers stunning views of the mountain from their shores. These lakes provide opportunities for boating, fishing, and picturesque walks where visitors can immerse themselves in the serenity of the landscape.

To enhance your visit to Mount Fuji, consider exploring the nearby cultural attractions as well. The Fuji-Hakone-Izu National Park, encompassing Mount Fuji and its surrounding regions, is a treasure trove of natural wonders, hot springs, and traditional Japanese ryokans. A visit to the Fuji-Q Highland amusement park, nestled at the base of the mountain, offers thrilling roller coaster rides and breathtaking views of Fuji-san.

In conclusion, Mount Fuji's roaring beauty and cultural significance make it an unmissable destination for those seeking to experience the power and mystique of this iconic volcano. Whether climbing to the summit, marveling at its reflection in the Fuji Five Lakes, or immersing oneself in the surrounding natural and cultural attractions, a journey to Mount Fuji promises a once-in-a-lifetime experience that will leave an indelible mark on your soul.

Nature's Fury: The Cataclysmic Eruption of Krakatoa

In the annals of volcanic history, few events have captivated the world's attention as intensely as the cataclysmic eruption of Krakatoa. This volcanic island, situated in the Sunda Strait between Java and Sumatra in present-day Indonesia, unleashed one of the most devastating and far-reaching eruptions ever recorded. In this chapter, we delve into the awe-inspiring power of Krakatoa, exploring its geological significance, the catastrophic eruption of 1883, and the intriguing legacy it has left behind.

Krakatoa, also known as Krakatau, was a composite volcano formed through a process of volcanic activity over thousands of years. It consisted of three main volcanic cones: Rakata, Danan, and Perboewatan. These cones rose from the depths of the ocean, creating an island that spanned roughly 5 kilometers (3.1 miles) in diameter. Beneath the surface, intense tectonic forces were at work, as the Indo-Australian Plate was being subducted beneath the Eurasian Plate, leading to the accumulation of magma that would eventually unleash its fury.

The eruption of Krakatoa in August 1883 was a cataclysmic event that shook the world. It began with a series of violent explosions that sent plumes of ash, gases, and volcanic debris high into the atmosphere. These explosions were heard thousands of kilometers away, with reports of the sound reaching as far as Perth, Australia, and the island of Rodrigues in the Indian Ocean. The force of the eruption was estimated to be equivalent to 200 megatons of TNT, making it one of the most powerful volcanic events in recorded history.

The consequences of the Krakatoa eruption were devastating. Enormous tsunamis, triggered by the collapse of the volcano's flanks into the sea, crashed upon the surrounding coastal areas. Waves as high as 30 meters (98 feet) swept across the Sunda Strait, engulfing entire villages and causing widespread destruction. The coastal towns of Merak and Anyer were obliterated, and the wave reached as far as the coast of South Africa, causing noticeable disturbances.

The eruption also had a profound impact on the global climate. The massive quantities of volcanic ash and sulfur dioxide injected into the stratosphere created a veil that encircled the Earth, blocking sunlight and causing a cooling effect. The following year, 1884, became known as the "Year Without a Summer," as temperatures dropped and agricultural crops failed in many parts of the world. The stunning sunsets painted with vibrant hues resulted from the scattering of sunlight by the volcanic particles in the atmosphere.

The legacy of Krakatoa's eruption extends beyond its immediate impact. The volcanic debris ejected into the atmosphere produced spectacular atmospheric phenomena. The scattering of sunlight by the volcanic ash particles caused vivid and prolonged sunsets, which inspired artists such as Edvard Munch to create renowned works like "The Scream." The eruption also fueled scientific interest in volcanology and the study of volcanic processes, leading to advances in our understanding of these extraordinary events.

For those interested in exploring the remnants of Krakatoa, caution must be exercised. The eruption of 1883 dramatically altered the topography of the island. What remained of the volcanic cones was largely submerged underwater, with only remnants visible above the sea surface. These remnants include the peaks of Rakata and Danan, which serve as a haunting reminder of the island's tumultuous past.

Visiting Krakatoa requires careful planning and a sense of adventure. Boat tours are available from nearby coastal towns, providing an opportunity to witness the remnants of the volcanic island up close. However, it is important to choose reputable tour operators who prioritize safety and adhere to regulations established to protect visitors and the fragile ecosystem.

In conclusion, the cataclysmic eruption of Krakatoa stands as a testament to the awe-inspiring power of nature. Its impact on the world, both in terms of the devastation caused and the scientific knowledge gained, is a testament to the enduring fascination and respect we hold for the forces that shape our planet. While visiting Krakatoa presents challenges, the opportunity to witness the remnants of this extraordinary event offers a profound glimpse into the raw power and beauty of nature's fury.

The Peak of Mysticism: Mount Kilimanjaro's Eternal Charm

In the heart of East Africa, where the vast African savannah meets the sky, Mount Kilimanjaro rises with a mystical allure that has captivated adventurers and nature enthusiasts for centuries. This iconic stratovolcano, standing proudly as the highest peak on the African continent, holds a timeless charm that beckons those seeking an extraordinary journey. In this chapter, we embark on an exploration of Mount Kilimanjaro, delving into its geological significance, cultural importance, and the enchanting experiences that await those who venture to its summit.

Mount Kilimanjaro, located in northeastern Tanzania near the border with Kenya, stands as an imposing presence, reaching an elevation of 5,895 meters (19,341 feet) above sea level. It is a dormant volcano consisting of three distinct volcanic cones: Kibo, Mawenzi, and Shira. Kibo, the highest cone, is home to Uhuru Peak, the ultimate goal for climbers and the highest point on Kilimanjaro.

The geological formation of Mount Kilimanjaro is linked to the tectonic activities of the East African Rift System, where the African Plate is splitting apart. Over millions of years, the movement of the Earth's crust and the accumulation of volcanic materials have shaped the mountain we see today. Kilimanjaro's iconic snow-capped summit, adorned with glaciers and perennial ice, stands as a dramatic contrast to the surrounding tropical landscapes.

The allure of Mount Kilimanjaro extends beyond its geological marvels. The mountain holds profound cultural and spiritual significance for the people of Tanzania, particularly the Chagga and Maasai tribes. Known as the "Roof of Africa," Kilimanjaro is believed to be a sacred abode, a place where gods and spirits dwell. Its majestic presence has inspired local folklore, legends, and rituals, adding a layer of mysticism to the mountain's allure.

For adventurers eager to conquer Kilimanjaro, several routes offer different experiences and challenges. The Marangu Route, often referred to as the "Coca-Cola Route," is the most popular and well-trodden path. It features huts for overnight stays, making it a more comfortable option for climbers. The Machame Route, also known as the "Whiskey Route," is more rugged and offers stunning vistas of the surrounding landscapes. Other routes, such as the Lemosho and Rongai routes, provide unique perspectives and quieter trails.

Climbing Mount Kilimanjaro requires careful planning, physical preparation, and proper acclimatization. Altitude sickness can be a concern, so climbers must take their time ascending, allowing their bodies to adjust to the decreasing oxygen levels. It is recommended to choose a reputable tour operator that prioritizes safety, provides experienced guides, and ensures proper equipment and support throughout the journey.

The climb itself is a transformative experience. As climbers traverse through different climatic zones, they encounter diverse ecosystems, from dense rainforests to alpine deserts. The journey showcases the astonishing biodiversity of Kilimanjaro, with unique plant species, such as the iconic giant groundsel and the resilient Kilimanjaro tree, dotting the landscapes.

Reaching the summit of Mount Kilimanjaro is an achievement that leaves a lasting impression. Standing atop Uhuru Peak at sunrise, surrounded by an otherworldly panorama of vast African plains, is a moment of triumph and awe. The sense of accomplishment and the realization of being on the highest point in Africa is truly unforgettable.

For those who prefer not to undertake the physically demanding climb, Mount Kilimanjaro can still be appreciated from various viewpoints in the surrounding region. The foothills of Kilimanjaro offer opportunities for wildlife safaris in national parks such as Amboseli in Kenya and Kilimanjaro National Park in Tanzania.

These safaris allow visitors to spot iconic African wildlife, including elephants, giraffes, zebras, and more, against the backdrop of the majestic mountain.

In conclusion, Mount Kilimanjaro's eternal charm lies in its mystical presence, its geological wonders, and its cultural significance. Whether standing on its summit or witnessing its grandeur from afar, the allure of Kilimanjaro offers a journey of self-discovery, natural marvels, and a deep connection to the soul of Africa.

The Sleeping Giant Awakens: Mount St. Helens' Explosive Rebirth

Nestled within the scenic beauty of the Pacific Northwest in the United States, Mount St. Helens emerged from obscurity to etch its name into history with a cataclysmic eruption in 1980. This dormant stratovolcano, with its distinctive conical shape and lush surroundings, underwent a transformative event that reshaped the landscape and provided valuable insights into the power of volcanic forces. In this chapter, we embark on a journey to explore the explosive rebirth of Mount St. Helens, uncovering its geological significance, the eruption of 1980, and the remarkable recovery that followed.

Mount St. Helens, located in Washington State, forms part of the Cascade Range, a chain of volcanos that stretches from northern California to British Columbia, Canada. It stands at an elevation of 2,550 meters (8,363 feet) and is known for its symmetrical cone shape, which was formed through layers of lava, ash, and volcanic debris over thousands of years. Despite its serene appearance, Mount St. Helens' history is marked by periods of volcanic activity, culminating in the devastating eruption of 1980.

The eruption of Mount St. Helens on May 18, 1980, ranks among the most significant volcanic events in recent history. It began with a massive landslide triggered by a magnitude 5.1 earthquake, which caused the northern flank of the volcano to collapse. This sudden release of pressure unleashed a powerful lateral blast, hurtling a high-speed avalanche of superheated gas, rock fragments, and volcanic ash across the landscape.

The lateral blast, traveling at speeds of over 670 miles per hour (1,080 kilometers per hour), swept away trees, demolished structures, and transformed the surrounding terrain into a barren wasteland. The intense heat generated by the blast incinerated everything in its path, leaving behind a devastated zone known as the "Blast Zone."

The eruption also unleashed a colossal plume of ash and volcanic gases that soared high into the atmosphere. The ash column rose to an altitude of more than 15 miles (24 kilometers) within minutes, spreading ash across vast areas and causing darkness in nearby cities. The ashfall impacted several states and even reached as far as Canada, affecting air travel, disrupting daily life, and leaving an indelible mark on the region's history.

The immediate aftermath of the eruption revealed a landscape transformed. The once lush forests and pristine lakes surrounding Mount St. Helens were replaced by a desolate expanse of gray ash and mudflows. However, amidst the devastation, a remarkable process of rebirth and ecological rejuvenation began.

The recovery of Mount St. Helens' ecosystem serves as a testament to the resilience of nature. Within weeks of the eruption, plants began to recolonize the barren terrain. First came the pioneer species, such as fireweed and lupine, which were able to establish themselves in the nutrient-rich volcanic soil. Over time, other plant species gradually took root, leading to the return of a diverse forest ecosystem.

The eruption of Mount St. Helens also provided valuable scientific insights into volcanic processes. It sparked a renewed interest in studying the dynamics of volcanic eruptions, the effects of volcanic ash on the environment, and the potential hazards associated with volcanic activity. Scientists closely monitored the recovery of the ecosystem, studying the succession of plant species and observing the return of wildlife to understand the resilience of nature in the face of such catastrophic events.

For visitors eager to explore Mount St. Helens, there are several ways to experience the remarkable rebirth and ongoing geological activity. The Mount St. Helens National Volcanic Monument, established to preserve the volcanic landscape and provide educational opportunities, offers a range of visitor centers, interpretive exhibits, and hiking trails. The Johnston Ridge Observatory provides panoramic views of the volcano's crater and the surrounding landscape, offering visitors a glimpse into the destructive power and subsequent recovery of Mount St. Helens.

Hiking opportunities abound in the area, with trails leading to viewpoints where the extent of the eruption's impact can be observed. The Ape Cave, a lava tube formed by previous volcanic activity, allows visitors to explore the unique underground geological formations. Guided tours and ranger-led programs provide educational insights into the volcano's history, geology, and ongoing monitoring efforts.

When visiting Mount St. Helens, it is important to be aware of safety considerations. Volcanic activity, albeit currently dormant, should be monitored, and visitors should adhere to any guidelines or restrictions in place. Proper hiking gear, including sturdy footwear and layered clothing, should be worn, as weather conditions can change rapidly. Visitors are also encouraged to carry water, sunscreen, and insect repellent, as they explore the diverse terrain and absorb the beauty of the recovering ecosystem.

In conclusion, Mount St. Helens' explosive rebirth serves as a powerful reminder of the dynamic forces that shape our planet. The eruption of 1980 left a profound impact on the surrounding landscape, yet it also sparked a remarkable process of recovery and scientific understanding. A visit to Mount St. Helens allows us to witness the raw power of nature, appreciate the resilience of life, and gain a deeper appreciation for the ongoing processes that continue to shape our world.

The Pillar of Smoke: Discovering Mount Etna's Enigmatic Nature

Rising majestically on the island of Sicily in Italy, Mount Etna stands as a towering testament to the power and beauty of volcanic forces. This enigmatic stratovolcano, shrouded in myths and legends, has captivated the imaginations of explorers, artists, and scientists throughout history. In this chapter, we embark on a journey to uncover the mysteries of Mount Etna, exploring its geological significance, the allure of its eruptions, and the unique experiences that await those who venture to its slopes.

Mount Etna, known as "Mongibello" in the local Sicilian dialect, is an active stratovolcano located on the eastern coast of Sicily. It is the tallest volcano in Europe, standing proudly at an elevation of 3,350 meters (10,990 feet) above sea level. Etna's dominance over the surrounding landscape is awe-inspiring, with its conical shape and vast volcanic slopes adorned with lush vegetation, vineyards, and ancient villages.

Geologically, Mount Etna is a complex and dynamic system. It is characterized by a central crater, known as the Voragine, and four summit craters: Bocca Nuova, Northeast Crater, Southeast Crater, and Voragine. These craters have been formed and modified over time by repeated volcanic activity, with eruptions occurring from both the summit and flanks of the volcano. Etna's eruptions are primarily effusive, characterized by the slow and steady flow of lava, although it has also experienced explosive eruptions.

One of the remarkable aspects of Mount Etna is its frequent eruptive activity. Etna has a long history of eruptions, with recorded activity dating back thousands of years. This volcanic activity has shaped the surrounding landscape and influenced the lives of those living in its shadow. The eruptions have resulted in the formation of numerous cones, craters, and lava flows, leaving a visible record of Etna's turbulent history.

The allure of Mount Etna's eruptions lies in their mesmerizing displays of fire and smoke. During eruptions, molten lava cascades down the volcano's slopes, illuminating the night sky with its incandescent glow. The sight of lava fountains, spewing ash clouds, and the powerful roar of volcanic activity create a spectacle that both fascinates and humbles those fortunate enough to witness it.

For those eager to explore Mount Etna, there are various ways to experience its enigmatic nature. Guided tours, led by knowledgeable local guides and volcanologists, offer a deeper understanding of the volcano's geology, history, and ongoing monitoring efforts. These tours can take visitors to the summit area, where they can witness the volcanic craters up close and appreciate the vastness of the volcanic landscape.

Hiking trails cater to different fitness levels, offering opportunities for both leisurely walks and more challenging ascents. The trails wind through ancient lava fields, pine forests, and lunar-like landscapes, showcasing the diverse ecosystems that have adapted to Etna's volatile environment. Along the way, hikers may encounter fumaroles, steam vents, and the remnants of past eruptions, providing glimpses into the volcano's dynamic nature.

Visiting Mount Etna requires attention to safety considerations. As an active volcano, eruptions can occur, and the volcano is closely monitored for any signs of activity. Visitors should follow the guidance of local authorities, stay informed about current volcanic conditions, and adhere to any restrictions or safety measures in place. It is advisable to wear suitable hiking gear, including sturdy footwear and layered clothing, as weather conditions can change rapidly at higher elevations.

To enhance the visit, exploring the surrounding region of Sicily is highly recommended. The nearby cities of Catania and Taormina offer opportunities to experience the rich culture, history, and cuisine of the island. The Sicilian cuisine, renowned for its flavors and fresh ingredients, provides a delightful culinary journey that should not be missed.

In conclusion, Mount Etna's enigmatic nature and powerful eruptions make it a captivating destination for adventurers and nature enthusiasts. Its imposing presence, unique geological features, and rich cultural heritage offer a multifaceted experience that combines scientific fascination with awe-inspiring natural beauty. A journey to Mount Etna unveils the secrets of this ancient pillar of smoke, inviting us to witness the primal forces that shape our planet and leave a lasting impression on our souls.

Flames on Ice: The Intriguing Saga of Eyjafjallajökull

Nestled within the rugged landscapes of Iceland, Eyjafjallajökull stands as a captivating symbol of the country's fiery nature. This stratovolcano, with its unpronounceable name and remarkable eruptions, has gained worldwide attention for its ability to disrupt air travel and command headlines. In this chapter, we embark on an exploration of Eyjafjallajökull, unraveling its geological significance, the dramatic eruption of 2010, and the unique experiences that await those who venture to witness its icy flames.

Eyjafjallajökull is located in southern Iceland, with its name meaning "Island Mountain Glacier." The volcano is part of a larger volcanic system that includes the nearby Katla volcano. Eyjafjallajökull is crowned by a glacier that covers its summit, adding an ethereal beauty to its already impressive stature. The volcano's history of eruptions spans thousands of years, with the most notable recent eruption occurring in 2010.

The eruption of Eyjafjallajökull in 2010 had a profound impact on global air travel. The eruption began on April 14 and continued for several weeks, sending plumes of volcanic ash and steam high into the atmosphere. The ash cloud, carried by prevailing winds, spread across Europe, causing widespread disruption and the temporary closure of airspace in many countries. It was one of the largest disruptions to air travel in modern history.

The unique nature of Eyjafjallajökull's eruption lies in its interaction with the overlying glacier. As the magma erupted from the volcano's vent, it encountered the ice and water within the glacier, leading to a violent phreatomagmatic eruption. The interaction between the magma and the water created explosive steam-driven eruptions, propelling ash and volcanic material into the sky.

The 2010 eruption of Eyjafjallajökull also showcased the resilience and adaptability of the Icelandic people. Evacuations were carried out swiftly and efficiently, ensuring the safety of residents in the surrounding areas. Iceland's robust monitoring and emergency response systems played a crucial role in mitigating the impact of the eruption and keeping the affected communities informed and prepared.

For those fascinated by the intriguing saga of Eyjafjallajökull, visiting the volcano and its surrounding areas provides a unique opportunity to witness the forces of nature at play. The region offers various hiking trails that allow visitors to explore the volcanic landscapes and gain a deeper understanding of the geology and impact of the eruption. Guided tours, led by knowledgeable local guides, offer insights into the volcano's history, the dynamics of its eruptions, and ongoing monitoring efforts.

It is important to note that accessing the summit of Eyjafjallajökull itself is not recommended due to the glacier's unpredictable nature and the potential dangers associated with volcanic activity. However, guided tours and hikes in the vicinity of the volcano provide an up-close and personal encounter with the surrounding landscapes and the remnants of past eruptions.

When visiting Eyjafjallajökull and its surrounding areas, it is crucial to prioritize safety. Iceland's weather conditions can be unpredictable, and the terrain can be challenging. It is advisable to come prepared with appropriate clothing, sturdy footwear, and outdoor equipment suitable for hiking in diverse conditions. It is also recommended to check weather forecasts, road conditions, and any travel advisories before embarking on the journey.

To enhance the visit, exploring the wider wonders of Iceland is highly recommended. The country boasts breathtaking landscapes, including cascading waterfalls, geothermal hot springs, and expansive lava fields. The vibrant capital city of Reykjavik offers a blend of modern amenities, cultural attractions, and a vibrant nightlife scene.

In conclusion, Eyjafjallajökull's intriguing saga showcases the interplay between fire and ice, offering a fascinating glimpse into the powerful forces that shape our world. Its eruption in 2010 serves as a reminder of the delicate balance between nature's beauty and its potential for disruption. A visit to Eyjafjallajökull and its surrounding areas grants us the opportunity to witness the harmonious dance of flames on ice, leaving us in awe of the wonders of Iceland's volcanic landscapes and the resilience of its people.

The Island Builder: Mauna Loa's Mighty Lava Flows

Amidst the vast expanse of the Pacific Ocean, the Hawaiian Islands rise with an elemental force. Among these volcanic wonders, Mauna Loa stands as a testament to the raw power and grandeur of nature. This colossal shield volcano, renowned for its mighty lava flows and astonishing size, has shaped the landscapes of Hawaii and continues to shape our understanding of the Earth's geology. In this chapter, we embark on a journey to explore the island-building prowess of Mauna Loa, uncovering its geological significance, its captivating lava flows, and the awe-inspiring experiences that await those who venture to its slopes.

Mauna Loa, whose name means "Long Mountain" in Hawaiian, is one of the five volcanoes that form the Big Island of Hawaii. It is the largest active volcano on Earth, with an estimated volume exceeding 75,000 cubic kilometers (18,000 cubic miles). Rising from the ocean floor, Mauna Loa reaches a towering height of 4,169 meters (13,678 feet) above sea level, making it a prominent feature of the Hawaiian archipelago.

The geological significance of Mauna Loa lies in its unique shield volcano structure. Unlike stratovolcanoes with steep slopes and explosive eruptions, shield volcanoes have a gentle, broad shape formed by successive layers of fluid lava flows. The lava, sourced from Mauna Loa's massive magma chamber beneath the surface, travels long distances, creating vast lava fields and adding to the volcano's immense size.

Mauna Loa has a rich volcanic history, with eruptions occurring at irregular intervals. Historical records indicate that the volcano has erupted 33 times since its first well-documented eruption in 1843. These eruptions have ranged from effusive lava flows to more explosive events, shaping the landscapes of the Big Island and leaving a remarkable legacy of volcanic activity.

The lava flows of Mauna Loa are a spectacle that showcases the volcano's island-building prowess. When an eruption occurs, rivers of molten lava cascade down the volcano's slopes, advancing at remarkable speeds and engulfing everything in their path. The lava, often described as "pahoehoe" when it forms smooth, undulating surfaces, creates intricate patterns and shapes as it cools and solidifies.

Visiting Mauna Loa offers a unique opportunity to witness the remarkable lava flows and explore the volcanic landscapes. Due to the volcano's remote and rugged nature, it is advisable to join guided tours or hikes led by experienced local guides. These tours provide insights into the geology, history, and cultural significance of Mauna Loa, offering a deeper understanding of the volcano's immense power.

It is important to note that exploring Mauna Loa requires careful planning and consideration for safety. The volcano is an active geological feature, and eruptions can occur. Visitors should stay informed about volcanic activity, adhere to any guidance or restrictions provided by local authorities, and follow the advice of experienced guides. It is essential to wear appropriate hiking gear, including sturdy footwear, sun protection, and layered clothing to prepare for changing weather conditions at higher elevations.

Beyond the volcanic landscapes, the Big Island of Hawaii offers a wealth of natural wonders and cultural experiences. The Hawai'i Volcanoes National Park, encompassing Mauna Loa and its sister volcano Kilauea, showcases the dynamic nature of volcanic activity in the region. Visitors can explore the park's hiking trails, witness active lava flows (when accessible and safe), and learn about the unique ecosystems that have adapted to volcanic environments.

Additionally, the Big Island boasts stunning beaches, lush rainforests, and a rich Polynesian heritage. Exploring the island's diverse landscapes, including the dramatic Waipi'o Valley and the breathtaking waterfalls of Akaka Falls State Park, provides a deeper appreciation for the beauty and diversity of Hawaii.

In conclusion, Mauna Loa's mighty lava flows and island-building prowess epitomize the powerful forces that shape the Hawaiian Islands. Its sheer size, geological significance, and captivating volcanic activity make it an incredible destination for adventurers and nature enthusiasts. A journey to Mauna Loa allows us to witness the ongoing processes that have shaped our world and experience the enchanting beauty of Hawaii's volcanic landscapes.

The Sacred Fire Mountain: Pele's Realm of Kīlauea

In the heart of the Hawaiian archipelago, a sacred fire mountain known as Kīlauea rises with an otherworldly beauty and a deep spiritual significance. This active shield volcano, revered as the realm of the Hawaiian goddess Pele, captivates the imagination and draws visitors from around the world to witness its fiery displays and experience its mystical allure. In this chapter, we embark on a journey to explore the sacred realm of Kīlauea, delving into its geological significance, the profound cultural connections, and the unforgettable experiences that await those who venture to its slopes.

Kīlauea, whose name means "spewing" or "much spreading" in Hawaiian, is one of the five volcanoes that form the Big Island of Hawaii. It is one of the world's most active volcanoes, with a history of eruptions spanning thousands of years. This shield volcano stands at an elevation of 1,247 meters (4,091 feet) above sea level, with its summit caldera, known as Halemaʻumaʻu, holding a special place in Hawaiian mythology and culture.

The geological significance of Kīlauea lies in its ongoing volcanic activity and its transformative impact on the landscape. Unlike many other volcanoes, Kīlauea primarily exhibits effusive eruptions, characterized by the continuous flow of fluid lava rather than explosive events. Lava fountains, lava lakes, and the creation of new land through lava flows have shaped the southeastern region of the Big Island, creating a dynamic and ever-changing landscape.

Kīlauea holds deep cultural and spiritual significance for the Hawaiian people. It is believed to be the dwelling place of Pele, the goddess of fire, lightning, wind, and volcanoes in Hawaiian mythology. Pele is both revered and feared as a powerful and unpredictable deity who shapes the land through her fiery manifestations. For the Hawaiian people, Kīlauea is a sacred realm, and paying homage to Pele is an integral part of their cultural identity.

Visiting Kīlauea offers a profound opportunity to connect with the volcano's spiritual and natural wonders. Hawai'i Volcanoes National Park, encompassing Kīlauea and its sister volcano Mauna Loa, provides a gateway to this extraordinary realm. The park offers a range of experiences, from scenic drives and overlooks to immersive hikes and educational exhibits.

One of the highlights of a visit to Kīlauea is witnessing the mesmerizing glow of the Halemaʻumaʻu crater at night. As darkness falls, the lava lake within the crater emits an ethereal light, casting an enchanting glow on the surrounding landscapes. The sight of lava spattering and the sound of its hissing offers a captivating display of the volcano's power and Pele's presence.

Hiking trails within the national park allow visitors to explore the diverse landscapes and witness the traces of past eruptions. The Crater Rim Trail offers stunning panoramic views of the volcano's caldera and the surrounding steam vents, while the Devastation Trail showcases the stark aftermath of Kīlauea's explosive eruptions. For a more immersive experience, guided tours and ranger-led programs provide insights into the volcano's geology, cultural significance, and ongoing monitoring efforts.

It is important to prioritize safety when visiting Kīlauea. Volcanic activity can be unpredictable, and conditions can change rapidly. Visitors should stay informed about current volcanic conditions, adhere to any park closures or restrictions, and follow the guidance of park rangers and local authorities. It is essential to come prepared with suitable hiking gear, including sturdy footwear, sun protection, and layered clothing to adapt to changing weather conditions.

Beyond Kīlauea, the Big Island of Hawaii offers a wealth of natural and cultural wonders to explore. From the lush rainforests of the Hamakua Coast to the stunning beaches of the Kona and Hilo regions, the island provides a diverse range of experiences for every traveler. Delving into the local cuisine, learning about traditional Hawaiian arts and crafts, and immersing oneself in the vibrant local communities enrich the overall journey.

In conclusion, Kīlauea's status as a sacred fire mountain and its ongoing volcanic activity make it a place of profound spiritual significance and natural wonder. A visit to this realm offers a unique opportunity to connect with the power of Pele and witness the ever-changing landscapes shaped by the volcano's fiery displays. Exploring Kīlauea and the surrounding areas allows us to appreciate the cultural heritage, the geological forces at play, and the deep-rooted connections between the Hawaiian people and their awe-inspiring natural surroundings.

A Tale of Destruction: Mount Pinatubo's Devastating Eruption

Nestled within the island of Luzon in the Philippines, Mount Pinatubo holds a haunting tale of destruction and resilience. This stratovolcano, once a slumbering giant, erupted with unimaginable force in 1991, wreaking havoc on its surroundings and leaving an indelible mark on the region. In this chapter, we delve into the devastating eruption of Mount Pinatubo, exploring its geological significance, the catastrophic events that unfolded, and the remarkable recovery that followed.

Mount Pinatubo, located in the Zambales Mountains, is part of the Pacific Ring of Fire—a region known for its volcanic activity. Prior to the eruption of 1991, the volcano had been dormant for over 600 years, leading many to believe it posed little threat. However, in a dramatic turn of events, Pinatubo awakened with catastrophic consequences.

The eruption of Mount Pinatubo began on June 9, 1991, following a series of seismic activity and precursor eruptions. The eruption was preceded by a massive explosion, known as a Plinian eruption, which sent a towering column of ash and volcanic gases high into the atmosphere. The eruption column reached an astonishing height of 40 kilometers (25 miles), darkening the skies and blanketing the surrounding areas in ash.

The eruption unleashed pyroclastic flows—devastating avalanches of hot gas, ash, and volcanic debris—that raced down the slopes of the volcano at incredible speeds. These pyroclastic flows and the subsequent lahars, or volcanic mudflows, caused extensive damage to the surrounding landscape, burying entire towns and displacing thousands of people. The impact of the eruption was felt far beyond the immediate vicinity, with ash and volcanic material reaching as far as neighboring countries and affecting global climate patterns.

The eruption of Mount Pinatubo is considered one of the most significant volcanic events of the 20th century. It had far-reaching environmental and socio-economic consequences. The eruption released massive amounts of sulfur dioxide into the atmosphere, forming a sulfuric acid haze that led to a significant cooling effect on the Earth's climate. The global temperatures were temporarily reduced, resulting in a noticeable impact on weather patterns and agricultural productivity.

In the aftermath of the eruption, a remarkable recovery effort took place. The affected communities rallied together, rebuilding their lives and infrastructure with resilience and determination. Efforts were made to mitigate the effects of lahars, including the construction of dikes and diversion channels to redirect future mudflows away from populated areas. The disaster also prompted the establishment of monitoring systems to better understand volcanic activity and enhance early warning systems for future eruptions.

Visiting Mount Pinatubo today offers an opportunity to witness the power of nature and the ongoing processes of recovery. Guided tours are available, providing visitors with insights into the geological history of the volcano, the impact of the eruption, and the resilience of the affected communities. These tours often involve a trek through the stunning landscapes and lava formations left behind by the eruption.

It is important to prioritize safety when visiting Mount Pinatubo. Volcanic activity can still occur, and conditions can change rapidly. It is recommended to join organized tours led by experienced local guides who are familiar with the area and adhere to safety protocols. Suitable hiking gear, including sturdy footwear, sun protection, and appropriate clothing, should be worn.

Exploring the surrounding region of Luzon also offers a wealth of cultural and natural wonders. The Philippines is renowned for its stunning beaches, vibrant cities, and warm hospitality. From the historic streets of Manila to the breathtaking rice terraces of Banaue, the country offers a diverse range of experiences that complement the visit to Mount Pinatubo.

In conclusion, Mount Pinatubo's devastating eruption serves as a stark reminder of the unpredictable power of volcanoes. The catastrophic events of 1991 left a lasting impact on the region and the world. However, the recovery and resilience of the affected communities exemplify the human spirit's ability to rebuild and overcome adversity. A visit to Mount Pinatubo allows us to witness the remnants of destruction, gain a deeper understanding of volcanic forces, and appreciate the strength and determination of the Filipino people.

Beyond the Clouds: Reaching the Summit of Mount Cotopaxi

Rising majestically in the Ecuadorian Andes, Mount Cotopaxi stands as a beacon of adventure and a symbol of natural beauty. This snow-capped stratovolcano, with its perfect cone shape and awe-inspiring presence, has long captivated the hearts of mountaineers and nature enthusiasts. In this chapter, we embark on a journey to explore the challenges and rewards of reaching the summit of Mount Cotopaxi, uncovering its geological significance, the allure of its icy peaks, and the indescribable experience that awaits those who dare to venture beyond the clouds.

Mount Cotopaxi, whose name means "neck of the moon" in the indigenous Quechua language, is one of the highest active volcanoes in the world. It soars to an elevation of 5,897 meters (19,347 feet) above sea level, making it a formidable challenge for mountaineers seeking to conquer its summit. The volcano is located within the Cotopaxi National Park, a protected area that encompasses a diverse range of ecosystems and serves as a sanctuary for unique flora and fauna.

Geologically, Cotopaxi is part of the Pacific Ring of Fire and is considered one of the most active volcanoes in Ecuador. Its symmetrical cone shape is the result of layers of volcanic ash, lava, and debris accumulated over thousands of years. Cotopaxi's last major eruption occurred in 1877, with subsequent smaller eruptions in the 20th century. The volcano's ongoing volcanic activity contributes to its mystical allure and adds an element of excitement for those who seek to conquer its icy slopes.

Reaching the summit of Mount Cotopaxi is an arduous but immensely rewarding endeavor. It requires physical fitness, mountaineering skills, and acclimatization to the high-altitude environment. Ascending the volcano involves traversing challenging terrain, including steep ice slopes, crevasses, and rocky ridges. Proper training and preparation, including altitude acclimatization and familiarity with mountaineering equipment and techniques, are crucial for a safe and successful climb.

Mountaineers attempting to summit Cotopaxi typically start their journey from the base camp, located at an altitude of approximately 4,800 meters (15,748 feet). From there, they navigate through a mix of snow, ice, and rocky terrain, utilizing crampons, ice axes, and ropes for safety and stability. The climb is a test of endurance, mental fortitude, and teamwork, as climbers push their limits in the face of extreme weather conditions and high-altitude challenges.

As climbers ascend higher, they enter a world beyond the clouds. The air becomes thinner, the landscape more rugged, and the vistas more breathtaking. The sight of the surrounding Andean peaks, glaciers, and the sprawling valleys below creates an otherworldly panorama that words cannot fully capture. It is a humbling experience that instills a deep sense of connection with the natural world and reminds climbers of their place within it. Reaching the summit of Mount Cotopaxi offers a triumphant sense of achievement and the chance to stand atop one of the world's most iconic volcanoes. The feeling of conquering the mountain, overcoming physical and mental challenges, and witnessing the beauty of the Ecuadorian Andes from such a vantage point is truly unforgettable. It is a testament to the indomitable human spirit and the pursuit of adventure.

For those who seek to experience Mount Cotopaxi's grandeur but may not be mountaineers, there are still opportunities to explore the lower reaches of the volcano. Cotopaxi National Park offers hiking trails, guided tours, and breathtaking viewpoints that allow visitors to immerse themselves in the unique natural landscapes surrounding the volcano. The park is home to diverse flora and fauna, including the Andean condor, llamas, and rare high-altitude plant species.

When visiting Mount Cotopaxi, it is important to prioritize safety and adhere to park regulations. Weather conditions at high altitudes can be unpredictable, and the volcano's volcanic activity should be closely monitored. Climbing with experienced guides who are familiar with the terrain and equipped with proper safety gear is recommended for mountaineering expeditions. For hikers, it is advisable to come prepared with appropriate clothing, sturdy footwear, and essentials such as water, sunscreen, and insect repellent. Exploring the surrounding region of Ecuador adds a rich cultural and natural dimension to the journey. The country offers vibrant markets, colonial cities, and diverse ecosystems, from the lush Amazon rainforest to the enchanting Galápagos Islands. Exploring the historical city of Quito, a UNESCO World Heritage site, or venturing into the pristine cloud forests and volcanic landscapes of Ecuador provides a well-rounded experience for travelers.

In conclusion, the journey to the summit of Mount Cotopaxi is a true test of strength, endurance, and determination. It offers a profound connection with the natural world, a breathtaking display of volcanic beauty, and a sense of accomplishment that lingers long after the descent. Whether conquering the summit or immersing oneself in the lower reaches of the volcano, a visit to Mount Cotopaxi is an unforgettable adventure that ignites the spirit of exploration and invites us to embrace the wonders of the Andean wilderness.

The Gateway to Hell: Mount Nyiragongo's Incandescent Crater

In the Democratic Republic of Congo, a natural wonder awaits those with a spirit of adventure and a fascination for the extraordinary. Mount Nyiragongo, with its mesmerizing incandescent crater, stands as a gateway to an infernal realm. This active stratovolcano, known for its relentless lava lake and awe-inspiring eruptions, captivates the imagination and beckons explorers to witness its fiery display. In this chapter, we embark on a journey to explore Mount Nyiragongo's geological significance, the enigmatic allure of its incandescent crater, and the extraordinary experiences that await those who dare to venture to this gateway to hell.

Mount Nyiragongo is located within Virunga National Park, a UNESCO World Heritage site in the eastern part of the Democratic Republic of Congo. It is one of the world's most active volcanoes and is renowned for its unique lava lake that resides within its crater. Standing at an elevation of 3,470 meters (11,380 feet), Nyiragongo's symmetrical shape and imposing presence dominate the surrounding landscape.

Geologically, Nyiragongo is classified as a stratovolcano, characterized by steep slopes and explosive eruptions. It is part of the East African Rift system, which stretches across several countries in the region. The volcano's eruptions are primarily of the effusive type, featuring highly fluid lava that cascades down the slopes with incredible speed and intensity.

The most captivating feature of Mount Nyiragongo is its lava lake. Nestled within the volcano's crater, this molten pool of fiery magma captivates onlookers with its mesmerizing glow and constant motion. The lava lake's incandescent display, often accompanied by fountains and spattering, creates a scene reminiscent of an otherworldly inferno. Its continuous activity and ever-changing patterns offer a unique spectacle that has attracted scientists and adventurers alike.

Visiting Mount Nyiragongo presents an extraordinary opportunity to witness the volcano's incandescent crater up close. A trek to the summit is an adventure of a lifetime, but it requires physical fitness, proper preparation, and adherence to safety guidelines. The hike typically begins from Kibati, a village located on the outskirts of Goma, the largest city in eastern Congo. From there, climbers embark on a challenging ascent through lush forests and volcanic terrain.

The hike to the summit of Mount Nyiragongo typically takes around five to six hours, depending on the pace and fitness level of the climbers. As the altitude increases, the air becomes thinner, and the hike becomes more demanding. It is essential to acclimatize, stay hydrated, and listen to the guidance of experienced guides who are familiar with the trail and safety protocols.

Reaching the summit rewards adventurers with an otherworldly view of the lava lake. The sight of the glowing magma, the sounds of its constant churning, and the heat emanating from the crater create an indescribable experience that leaves a lasting impression. Spending the night at the summit in basic but comfortable shelters allows visitors to marvel at the fiery spectacle under the cover of darkness—an unforgettable sight that etches itself into the memory.

While visiting Mount Nyiragongo, it is important to be mindful of safety considerations. Volcanic activity can be unpredictable, and conditions can change rapidly. Adhering to the guidance of experienced guides and park authorities is crucial for a safe and enjoyable visit. It is essential to come prepared with suitable hiking gear, including sturdy footwear, warm clothing for the summit, and camping essentials.

Exploring the wider Virunga National Park adds to the richness of the journey. The park is home to a remarkable diversity of wildlife, including the endangered mountain gorillas. Guided treks to encounter these gentle giants offer a unique and humbling wildlife experience. Additionally, the park's lush forests, volcanic landscapes, and pristine lakes contribute to its status as a natural treasure of immense value.

In conclusion, Mount Nyiragongo's incandescent crater serves as a gateway to an extraordinary realm of fire and spectacle. A visit to this remarkable volcano presents an opportunity to witness the captivating lava lake and immerse oneself in the natural wonders of the surrounding Virunga National Park. It is a journey that challenges the limits of human exploration while inspiring a deep appreciation for the powerful forces that shape our planet. The experience of standing before the gateway to hell is one that lingers in the hearts and minds of adventurers forever.

The Jewel of the Caribbean: Soufrière Hills and the Destruction of Montserrat

In the turquoise waters of the Caribbean Sea lies the picturesque island of Montserrat, known as the "Emerald Isle of the Caribbean" for its lush green landscapes and captivating beauty. However, beneath the idyllic surface lies a tale of devastation and resilience. The Soufrière Hills volcano, located on this enchanting island, erupted with unprecedented force, forever altering the lives of its inhabitants and leaving an indelible mark on the landscape. In this chapter, we delve into the geological significance of Soufrière Hills, the catastrophic events that unfolded, and the remarkable spirit of the Montserratian people who have rebuilt their lives amidst the destruction.

Soufrière Hills, a stratovolcano, lies in the southern part of Montserrat. Prior to its reawakening in 1995, the volcano had been dormant for over 400 years, with no historical records of eruptions. However, a series of seismic activity and ash emissions in 1995 marked the beginning of a catastrophic chapter in Montserrat's history.

The eruption of Soufrière Hills commenced on July 18, 1995, and unleashed a series of pyroclastic flows and explosive eruptions that shook the island to its core. These pyroclastic flows, consisting of hot gas, ash, and volcanic debris, rushed down the slopes of the volcano at astonishing speeds, obliterating everything in their path. The destruction caused by the eruption was compounded by lahars, or volcanic mudflows, triggered by heavy rainfall mixing with the loose volcanic material.

The impact of the eruption was devastating. The capital city of Plymouth, once a bustling hub of activity, was buried under layers of ash and volcanic debris, rendering it uninhabitable. More than half of the island's population was forced to evacuate, seeking refuge in other parts of Montserrat and abroad. The destruction of homes, infrastructure, and livelihoods left the Montserratian people facing an uncertain future.

The eruption of Soufrière Hills is an ongoing event, with intermittent periods of activity and relative calm. Volcanologists and scientists closely monitor the volcano, studying its behavior and providing early warning systems to mitigate risks. The volcano's activity has shaped the landscape of Montserrat, with new lava domes forming and altering the island's topography over time.

Despite the devastation, the people of Montserrat have shown remarkable resilience and determination in rebuilding their lives. Efforts to relocate the capital to the northern part of the island, away from the volcanic hazards, have been underway, with the construction of new infrastructure and the revitalization of communities. The Montserrat Volcano Observatory provides vital monitoring and research, keeping the island's residents informed and prepared.

Visiting Montserrat offers a unique opportunity to witness the impacts of a volcanic eruption and the resilience of a community. The exclusion zone around the volcano restricts access to certain areas, including the former capital of Plymouth, due to ongoing volcanic activity and safety concerns. However, guided tours and observation points allow visitors to witness the remnants of the destruction and gain insights into the geological processes at play.

Exploring the northern part of Montserrat provides an opportunity to experience the island's natural beauty and cultural heritage. The Montserrat Cultural Centre in Little Bay showcases the island's rich history, music, and art, while hiking trails lead to scenic viewpoints that offer panoramic vistas of the Caribbean Sea. The island's pristine beaches, such as Rendezvous Bay and Little Bay Beach, provide opportunities for relaxation and rejuvenation.

It is essential to prioritize safety when visiting Montserrat. Volcanic activity can be unpredictable, and conditions can change rapidly. Visitors should heed the guidance of local authorities, adhere to safety restrictions, and stay informed about the current status of Soufrière Hills. It is also recommended to engage the services of knowledgeable local guides who can provide insights into the volcano's history and ongoing activity.

In conclusion, the story of Soufrière Hills and the destruction of Montserrat is a testament to the forces of nature and the resilience of the human spirit. The eruption serves as a reminder of the ever-present risks posed by volcanic activity, but it also highlights the indomitable will of a community to rebuild and create a brighter future. A visit to Montserrat offers a chance to witness the island's natural beauty, learn from its tumultuous history, and embrace the spirit of hope and resilience that continues to define this jewel of the Caribbean.

A Charming Fury: Mount Agung's Volcanic Ballet

Amidst the lush landscapes and vibrant culture of Bali, Indonesia, a formidable presence rises in the form of Mount Agung. This stratovolcano, with its graceful slopes and spiritual significance, showcases the delicate dance between beauty and fury. Mount Agung, known as the "Mother Mountain" to the Balinese people, holds a captivating allure, drawing visitors from around the world to witness its volcanic ballet. In this chapter, we embark on a journey to explore the geological significance of Mount Agung, the enigmatic charm of its eruptions, and the unforgettable experiences that await those who embrace this charming fury.

Mount Agung, standing tall at an elevation of 3,031 meters (9,944 feet) above sea level, is located in the eastern part of Bali. It is the highest point on the island and holds immense cultural and spiritual significance to the Balinese people. Known as the abode of the gods, Mount Agung is believed to be a sacred place where deities reside, playing a central role in the island's religious and cultural practices.

Geologically, Mount Agung is classified as an active stratovolcano, formed by layers of volcanic ash, lava, and rock. Its symmetrical cone shape is the result of multiple eruptions over thousands of years. The volcano's last major eruption occurred in 1963, with subsequent eruptions and volcanic activity in the years that followed.

The eruptions of Mount Agung are a spectacle that blends power and beauty in a captivating ballet of nature. The volcano's eruptions are often characterized by towering ash plumes, glowing lava flows, and volcanic bombs launched into the air. These eruptions showcase the dynamic nature of Mount Agung and remind us of the primal forces that shape our planet.

Visiting Mount Agung offers a unique opportunity to witness the volcano's volcanic ballet. Trekking to the summit requires physical fitness, proper preparation, and adherence to safety guidelines. The ascent typically begins from the village of Besakih, known as the "Mother Temple" and considered the holiest temple in Bali. From there, climbers navigate through lush forests, rocky terrain, and volcanic landscapes, guided by experienced local mountaineers.

Reaching the summit of Mount Agung rewards adventurers with panoramic views of Bali's stunning landscapes. The sight of the surrounding mountains, including Mount Rinjani on the neighboring island of Lombok, and the vast expanse of the Bali Sea creates a sense of awe and wonder. Witnessing the sunrise from the summit adds an ethereal touch to the experience, as the first rays of light illuminate the island and cast a golden glow on the volcanic slopes.

While Mount Agung's eruptions are a mesmerizing display of nature's power, it is essential to prioritize safety when visiting. The volcano is an active geological feature, and volcanic activity can be unpredictable. Staying informed about the volcano's status and adhering to safety guidelines issued by local authorities and guides is crucial. Engaging the services of experienced guides who are familiar with the area and equipped with proper safety gear is recommended for trekking expeditions.

Exploring the wider island of Bali provides a wealth of cultural and natural wonders to complement the visit to Mount Agung. From the vibrant arts scene and traditional ceremonies in Ubud to the stunning rice terraces of Jatiluwih, Bali offers a rich tapestry of experiences. Visitors can immerse themselves in the island's unique culture, indulge in traditional cuisine, and relax on pristine beaches such as Kuta and Nusa Dua.

In conclusion, Mount Agung's volcanic ballet showcases the harmonious interplay between beauty and fury. A visit to this captivating volcano allows us to witness the delicate dance of its eruptions, embrace its cultural and spiritual significance, and connect with the natural wonders of Bali. It is a journey that instills a sense of reverence for the primal forces that shape our world, reminding us of the enchanting charm and untamed power of Mount Agung.

The Frozen Inferno: Mount Erebus and Antarctica's Fiery Secret

In the vast expanse of the frozen continent of Antarctica, a hidden fiery secret lies beneath the icy surface. Mount Erebus, a mesmerizing stratovolcano, stands as a testament to the juxtaposition of ice and fire in the world's coldest and most remote region. This enigmatic volcano, with its active lava lake and icy summit, captivates the imagination and beckons explorers to uncover its frozen inferno. In this chapter, we embark on a journey to explore the geological significance of Mount Erebus, the unique characteristics of its volcanic activity, and the extraordinary experiences that await those who dare to venture to this icy realm.

Mount Erebus is situated on Ross Island, near the continent's southernmost point. It is the southernmost active volcano on Earth and reaches an impressive height of 3,794 meters (12,448 feet) above sea level. Erebus is part of the Pacific Ring of Fire, a region known for its tectonic activity and volcanic eruptions.

Geologically, Mount Erebus is classified as a stratovolcano, characterized by its steep slopes and explosive eruptions. What sets Erebus apart is the presence of a persistent lava lake within its summit crater—the only known lava lake on the continent. This unique feature makes Erebus a subject of great interest to volcanologists and scientists who study the volcano's behavior and the implications of its volcanic activity in Antarctica.

The lava lake of Mount Erebus is a mesmerizing sight to behold. With its ever-present glow and constantly shifting patterns, it offers a glimpse into the fiery heart of the volcano. The lava lake's activity is known to vary, with occasional eruptions and changes in the lava's behavior and composition. Its persistent presence challenges the conventional understanding of volcanic activity in extreme cold environments.

Visiting Mount Erebus is a rare opportunity to witness the frozen inferno and explore the breathtaking landscapes of Antarctica. Due to the extreme remoteness and harsh conditions of the region, visiting Mount Erebus requires careful planning, specialized equipment, and guidance from experienced expedition leaders.

One of the primary ways to visit Mount Erebus is through scientific research expeditions and specialized tour operators that offer guided trips to Antarctica. These trips often involve traveling by ship or plane to reach the continent, followed by helicopter transfers to Ross Island, where Erebus is located. Travelers should ensure that the tour operator is reputable, with a strong focus on environmental conservation and adherence to strict guidelines to minimize human impact on the fragile Antarctic ecosystem.

Upon arrival at Ross Island, visitors can experience the grandeur of Mount Erebus through guided tours and hikes. These tours provide insights into the geological features of the volcano, the scientific research being conducted, and the unique challenges of studying an active volcano in such extreme conditions. It is important to follow the guidance of the expedition leaders, as safety is paramount in this harsh environment.

Although the ascent to the summit of Mount Erebus is reserved for trained scientists and mountaineers, the surrounding areas of Ross Island offer opportunities for exploration and appreciation of Antarctica's pristine beauty. Visitors can encounter diverse wildlife, such as seals and penguins, and marvel at the towering glaciers and ice formations that characterize the region.

As with any visit to Antarctica, it is essential to prioritize environmental responsibility and conservation. Travelers should follow the principles of Leave No Trace, respecting the fragile ecosystem and wildlife. Proper waste management, adherence to designated trails, and compliance with regulations set forth by international agreements, such as the Antarctic Treaty System, are crucial for preserving the integrity of this unique environment.

In conclusion, Mount Erebus stands as a frozen inferno—a remarkable testament to the remarkable forces that shape our planet, even in the harshest of environments. A visit to this enigmatic volcano offers a rare glimpse into the fiery heart of Antarctica, connecting us with the raw power of nature and the beauty of the icy wilderness. It is an expedition of awe and discovery, reminding us of the immense wonders that await those who dare to venture to the ends of the Earth.

The Sleeping Giant of Europe: Mount Teide's Majestic Silence

Amidst the stunning landscapes of the Canary Islands, a sleeping giant lies in peaceful slumber. Mount Teide, with its towering presence and majestic silence, stands as a symbol of Tenerife's natural beauty and geological marvels. This stratovolcano, the highest peak in Spain and one of the world's most impressive volcanic structures, captivates the hearts of visitors and beckons them to explore its ancient history and breathtaking vistas. In this chapter, we embark on a journey to unravel the geological significance of Mount Teide, embrace its tranquil power, and discover the extraordinary experiences that await those who venture to this majestic giant of Europe.

Mount Teide, located in Tenerife, the largest of the Canary Islands, rises to an elevation of 3,718 meters (12,198 feet) above sea level. It is a stratovolcano formed by successive volcanic eruptions over millions of years, resulting in its distinctive cone shape. The volcano is part of Teide National Park, a UNESCO World Heritage site that spans over 18,000 hectares and encompasses a diverse range of ecosystems.

Geologically, Mount Teide is an active volcano, although it has been dormant since its last eruption in 1909. It is considered one of the most important and best-studied volcanoes in the world, attracting scientists and researchers from various disciplines. The volcano's formation is linked to the complex tectonic processes occurring in the region, where the African and Eurasian tectonic plates converge.

Mount Teide's majestic silence is a testament to its current state of dormancy. However, its geological significance is evident in the diverse volcanic features that shape the landscape. From the striking caldera, Las Cañadas, which encircles the summit, to the rugged lava flows and intricate rock formations, Mount Teide offers a captivating geological tapestry for exploration.

Visiting Mount Teide presents a unique opportunity to immerse oneself in the natural wonders of Tenerife and witness the grandeur of this sleeping giant. The ascent to the summit is a popular activity for adventurers and nature enthusiasts, but it requires proper planning, permits, and adherence to safety guidelines.

To reach the summit of Mount Teide, visitors can choose from several hiking trails of varying difficulty. The most common trail starts from the base station at Montaña Blanca and leads to the summit through a well-marked path. The hike offers breathtaking views of the surrounding landscape, including the expansive Las Cañadas caldera and the neighboring islands. It is essential to be adequately prepared, with suitable clothing, sturdy footwear, and sufficient water and provisions.

For those who prefer a more leisurely experience, the Teide Cable Car provides an alternative means to reach a panoramic viewpoint near the summit. The cable car ride offers stunning vistas of the volcanic landscapes and allows visitors to enjoy the beauty of Mount Teide without the physical exertion of a long hike. It is important to note that the cable car operates subject to weather conditions and may require advance reservations during peak seasons.

Exploring Teide National Park beyond the summit area reveals a wealth of natural wonders. The park is home to unique flora and fauna adapted to the volcanic environment, including the endemic Teide violet and the Teide lizard. Hiking trails of various lengths and difficulties crisscross the park, allowing visitors to discover the diverse ecosystems, volcanic rock formations, and picturesque viewpoints.

When visiting Mount Teide, it is essential to prioritize safety and environmental responsibility. Weather conditions at high altitudes can be unpredictable, and sudden changes may occur. It is advisable to check weather forecasts, follow the guidance of park authorities, and be prepared for varying temperatures and strong winds.

Furthermore, respecting the natural environment and adhering to park regulations are crucial for the preservation of Teide National Park. Visitors should stay on designated trails, avoid littering, and refrain from damaging or removing any natural or cultural artifacts. By being responsible visitors, we can contribute to the long-term conservation of this remarkable natural treasure.

In conclusion, Mount Teide's majestic silence echoes its dormant state and reveals the splendor of Tenerife's volcanic heritage. A visit to this sleeping giant of Europe offers a unique opportunity to explore its ancient geological formations, embrace the tranquility of its surroundings, and witness the awe-inspiring vistas that stretch across the Canary Islands. It is a journey that inspires a deep connection with nature, ignites the spirit of exploration, and leaves an indelible impression of Mount Teide's timeless allure.

The Great Pacific Fireworks: Exploring Popocatépetl's Explosive Show

In the heart of Mexico, a fiery spectacle unfolds as one of the country's most active volcanoes takes center stage. Popocatépetl, with its towering presence and explosive temperament, commands attention and fascination. This stratovolcano, known as "Popo" by the locals, treats onlookers to a breathtaking display of pyrotechnics and the raw power of nature. In this chapter, we embark on a journey to explore the geological significance of Popocatépetl, unravel its explosive nature, and discover the extraordinary experiences that await those who venture to witness its great Pacific fireworks.

Popocatépetl is located in the Trans-Mexican Volcanic Belt, a region renowned for its volcanic activity. Standing tall at an elevation of 5,426 meters (17,802 feet), it is the second highest peak in Mexico and one of the most iconic and active volcanoes in the country. Popocatépetl's name translates to "Smoking Mountain" in the indigenous Nahuatl language, a fitting description for its frequent emissions of steam and ash.

Geologically, Popocatépetl is a stratovolcano, formed by successive layers of volcanic ash, lava, and debris. It is part of the Pacific Ring of Fire, a belt of intense tectonic activity encircling the Pacific

Ocean. This volcanic belt extends from the Americas to Asia and is responsible for a significant portion of the world's volcanic eruptions and seismic activity.

Popocatépetl's explosive nature is evident in its historical and ongoing volcanic activity. Throughout its recorded history, the volcano has experienced numerous eruptions, with varying degrees of intensity. Eruptions can include the emission of ash clouds, pyroclastic flows (dense mixtures of hot gas, ash, and rock fragments), and even the release of glowing lava bombs. The volcanic activity of Popocatépetl serves as a reminder of the ever-present dynamic forces shaping our planet.

Visiting Popocatépetl offers a unique opportunity to witness the great Pacific fireworks and experience the captivating power of this iconic volcano. It is important to note that due to its ongoing volcanic activity, access to the summit is restricted for safety reasons. However, there are still ways to appreciate and explore the area surrounding the volcano.

One popular option is to visit the nearby city of Puebla, located approximately 70 kilometers (43 miles) southeast of Popocatépetl. Puebla is a UNESCO World Heritage site renowned for its colonial architecture, vibrant culture, and rich history. From Puebla, visitors can enjoy panoramic views of the volcano, especially from elevated viewpoints such as the church of San Francisco Acatepec or the nearby town of Cholula, known for its ancient pyramid and stunning vistas.

For those seeking a closer encounter with Popocatépetl, guided tours and hikes are available to explore the lower reaches of the volcano. These excursions provide opportunities to witness the surrounding volcanic landscapes, including volcanic ash fields and the remnants of past eruptions. It is essential to embark on these activities with experienced guides who are knowledgeable about the area, safety protocols, and the latest information on the volcano's activity.

When visiting Popocatépetl, it is crucial to prioritize safety and stay informed about the volcano's current status. Monitoring reports from the National Center for Disaster Prevention (CENAPRED) provide up-to-date information on volcanic activity, allowing visitors to make informed decisions about their itinerary and any potential restrictions or advisories in place.

Furthermore, respecting the natural environment and local communities is essential. Visitors should adhere to park regulations, follow designated trails, and refrain from littering or damaging the delicate ecosystems surrounding the volcano. By practicing responsible tourism, we can ensure the preservation of this natural wonder for future generations.

In conclusion, Popocatépetl's explosive show captivates and astonishes, revealing the immense power and beauty of nature's forces. A visit to this great Pacific fireworks spectacle offers an opportunity to appreciate the geological significance of the volcano, witness its ongoing activity, and explore the surrounding volcanic landscapes. It is a journey that immerses us in the raw energy of Popocatépetl and leaves us in awe of the extraordinary volcanic wonders that shape our world.

In the Shadows of Giants: Hekla's Mysterious and Frequent Eruptions

In the rugged landscapes of Iceland, a mysterious giant looms, shrouded in a history of frequent eruptions and enigmatic allure. Hekla, known as the "Gateway to Hell" in Icelandic folklore, stands as one of the country's most active and captivating volcanoes. Its tumultuous past, characterized by frequent eruptions and towering plumes of ash, has both fascinated and instilled a sense of awe in those who venture to witness its explosive displays. In this chapter, we delve into the geological significance of Hekla, unravel its mysterious nature, and discover the extraordinary experiences that await those who dare to explore the shadows of this Icelandic giant.

Hekla, located in the southern part of Iceland, is a stratovolcano that stretches across a distance of approximately 25 kilometers (15.5 miles). Standing at an elevation of 1,491 meters (4,892 feet), it dominates the surrounding landscape, its steep slopes cloaked in a mix of snow and volcanic ash. Hekla's name derives from the Old Norse word "hekl," meaning "hooded," which aptly describes the volcano's appearance, often veiled in clouds and mist.

Geologically, Hekla is part of the larger East Volcanic Zone, which runs across the southern part of Iceland. It is situated on the Mid-

Atlantic Ridge, a tectonic boundary where the North American and Eurasian plates diverge. Hekla's frequent eruptions are attributed to the ongoing movement and interaction of these tectonic plates, creating a hotspot for volcanic activity.

Hekla's eruptive history is as captivating as it is mysterious. Over the centuries, the volcano has experienced numerous eruptions, earning a reputation for its unpredictable nature and powerful explosions. Hekla's eruptions can vary in intensity and eruption style, ranging from explosive to effusive, with lava flows streaming down its slopes. Despite its frequent eruptions, the volcano often provides signs of impending activity, including increased seismic activity and ground deformation, allowing scientists to monitor and study its behavior.

Visiting Hekla offers a unique opportunity to witness the dynamic nature of Iceland's volcanic landscape and immerse oneself in the geological wonders of the region. However, due to the potential hazards associated with volcanic activity, including ash clouds and lava flows, access to the summit area may be restricted during periods of increased volcanic activity. It is essential to stay informed about the volcano's current status and adhere to the guidance provided by local authorities and experts.

For those seeking to explore the surroundings of Hekla, several hiking trails offer a chance to experience the volcano's captivating presence. The most popular trail is the Hekla Summit Trail, which allows hikers to ascend to the summit and enjoy panoramic views of the surrounding landscapes. This hike requires a good level of fitness, proper equipment, and knowledge of the route, as weather conditions can change rapidly, and the terrain can be challenging.

Exploring the Hekla region beyond the volcano itself unveils a wealth of natural wonders. The surrounding landscape is marked by vast lava fields, steaming hot springs, and unique geological formations. The nearby Þjórsárdalur Valley offers picturesque landscapes, with rolling hills, waterfalls, and historical sites. Visitors can also explore the nearby Landmannalaugar region, renowned for its colorful rhyolite mountains, natural hot springs, and diverse hiking trails.

When visiting Hekla and its surroundings, it is important to prioritize safety and responsible tourism. Weather conditions in Iceland can be unpredictable, and proper preparation is essential. Hikers should equip themselves with suitable clothing, footwear, and provisions, and consider traveling with experienced guides who are knowledgeable about the area and equipped to handle potential challenges.

Respecting the natural environment is also crucial. Visitors should stay on designated trails, avoid damaging or removing any natural or cultural artifacts, and follow the principles of Leave No Trace. By minimizing our impact, we can help preserve the unique beauty and ecological integrity of the Hekla region.

In conclusion, Hekla's mysterious and frequent eruptions enthrall and mystify, revealing the raw power and beauty of Iceland's volcanic heritage. A visit to the shadows of this Icelandic giant allows us to immerse ourselves in the geological wonders of the region, witness the dynamic forces that shape our planet, and embrace the extraordinary landscapes that define this land of fire and ice. It is a journey that awakens the senses, sparks curiosity, and leaves an indelible mark on our understanding of nature's captivating power.

The Ring of Fire's Wrath: Unraveling the Tale of Mount Rainier

In the majestic landscapes of the Pacific Northwest, a towering giant stands guard, with its snow-capped peaks and awe-inspiring presence. Mount Rainier, an iconic stratovolcano, captures the imagination and reverence of all who behold it. This volcanic behemoth, nestled within the Cascade Range, holds a tumultuous past and an ongoing potential for volcanic activity. In this chapter, we embark on a journey to unravel the geological significance of Mount Rainier, delve into its captivating tale, and discover the extraordinary experiences that await those who venture to witness the Ring of Fire's wrath.

Mount Rainier, standing at an imposing elevation of 4,392 meters (14,411 feet), is the highest peak in Washington State and one of the most prominent volcanoes in the contiguous United States. It is located in Mount Rainier National Park, a designated UNESCO World Heritage site that spans over 369 square miles (956 square kilometers) of pristine wilderness.

Geologically, Mount Rainier is part of the Pacific Ring of Fire, a zone of intense tectonic activity encircling the Pacific Ocean. The volcano's formation can be traced back millions of years to the subduction zone where the Juan de Fuca plate is being forced beneath the North American plate. This ongoing tectonic interaction has created a hotspot for volcanic activity, resulting in the formation of the Cascade Range, of which Mount Rainier is a prominent member.

Mount Rainier's tumultuous past is written in its geological layers and glacial features. Over its extensive history, the volcano has experienced numerous eruptions, lahars (volcanic mudflows), and the formation of extensive glaciers. The most recent major eruption occurred around 1,000 years ago, resulting in the deposition of volcanic ash and pyroclastic flows that shaped the surrounding landscape.

Today, Mount Rainier remains an active volcano, classified as dormant rather than extinct. While no eruptions have been recorded in recent history, the potential for future volcanic activity remains. Ongoing monitoring efforts, including seismic activity and gas emissions, help scientists assess the volcano's behavior and potential hazards. Mount Rainier's status as a potentially active volcano underscores the need for preparedness and a comprehensive understanding of the region's volcanic hazards.

Visiting Mount Rainier offers a profound opportunity to witness the power and beauty of nature in a spectacular alpine setting. The national park surrounding the volcano provides a range of experiences for visitors. From leisurely strolls through old-growth forests to challenging alpine hikes and mountaineering expeditions, there is something for everyone.

One popular activity is the Wonderland Trail, a 93-mile (150-kilometer) loop that circumnavigates Mount Rainier. This multi-day trek offers panoramic views of the volcano, traverses diverse ecosystems, and showcases the park's natural wonders. Hiking the trail requires careful planning, adequate physical fitness, and obtaining necessary permits.

For those seeking a more relaxed experience, scenic drives along the Paradise and Sunrise visitor centers provide access to breathtaking vistas, alpine meadows adorned with wildflowers, and cascading waterfalls. These areas also offer opportunities for picnicking, wildlife viewing, and photography.

Guided tours and ranger-led programs are available to enhance visitors' understanding of the volcano's geology, ecology, and cultural significance. These educational opportunities provide insights into the region's volcanic history, the impacts of glacial activity, and the adaptations of plants and animals to the harsh mountain environment.

When visiting Mount Rainier, it is crucial to prioritize safety and preparedness. The weather in the high alpine environment can change rapidly, with the potential for severe storms even during summer months. Hikers and climbers should be equipped with appropriate gear, including sturdy footwear, warm clothing, and navigational tools. Additionally, it is important to stay informed about current trail conditions, adhere to park regulations, and carry out any waste in a Leave No Trace manner.

In conclusion, Mount Rainier's commanding presence and volcanic legacy invite exploration and reverence. A journey to this iconic volcano unveils a tale of geologic forces, glacial beauty, and the ever-present potential for volcanic activity. It is an opportunity to witness the Ring of Fire's wrath firsthand, to appreciate the wonders of Mount Rainier National Park, and to connect with the enduring power and grace of one of the Pacific Northwest's most captivating natural wonders.

The Island's Sentinel: Mount Yasur's Mesmerizing Lava Display

In the remote reaches of the South Pacific, a fiery sentinel stands guard over an island paradise. Mount Yasur, on the island of Tanna in Vanuatu, captivates the hearts and imaginations of all who witness its mesmerizing lava display. This stratovolcano, known as the "Lighthouse of the Pacific," treats visitors to a breathtaking spectacle of eruptive activity, offering a unique glimpse into the raw power and beauty of nature. In this chapter, we embark on a journey to explore the geological significance of Mount Yasur, unravel its captivating tale, and discover the extraordinary experiences that await those who venture to witness its mesmerizing lava display.

Mount Yasur, located on the eastern side of Tanna Island, rises to an elevation of 361 meters (1,184 feet) above sea level. Its prominent position and explosive nature have made it one of the most accessible and active volcanoes in the world. Mount Yasur is part of the Pacific Ring of Fire, a region characterized by intense volcanic and seismic activity due to the collision of tectonic plates.

Geologically, Mount Yasur is a stratovolcano formed by the accumulation of layers of volcanic ash, lava, and other ejecta. Its distinct cone shape is the result of repeated eruptions over time. The volcano's activity is fueled by the subduction of the Australian plate beneath the Pacific plate, creating a hotspot for volcanic activity in the region.

What sets Mount Yasur apart is its persistent Strombolian activity, characterized by frequent explosive eruptions and the ejection of incandescent volcanic bombs. This type of volcanic activity produces regular bursts of ash, gas, and lava, creating a mesmerizing lava display that draws visitors from around the world.

Visiting Mount Yasur offers a rare opportunity to witness the power and beauty of an active volcano up close. However, it is essential to prioritize safety and follow the guidance of experienced guides and local authorities. Mount Yasur is known for its ongoing volcanic activity, which can present hazards to visitors.

Guided tours and excursions are available to explore Mount Yasur's volcanic wonders. These tours provide insights into the volcano's geology, the cultural significance it holds for the local indigenous communities, and safety protocols to ensure a memorable and secure experience.

When visiting Mount Yasur, it is important to come prepared with suitable clothing, including sturdy footwear, as the terrain can be rough and rocky. The volcanic environment can be hot and dusty, so wearing protective clothing and bringing ample water is essential.

Due to the volcano's accessibility and popularity, it is advisable to visit Mount Yasur during non-peak hours or in the shoulder seasons to avoid overcrowding and ensure a more intimate experience. It is also recommended to check the volcano's activity level and consult with local authorities for any safety advisories or restrictions in place.

Experiencing Mount Yasur's mesmerizing lava display is a unique opportunity to witness the dynamic forces that shape our planet. It serves as a powerful reminder of the extraordinary geological wonders that exist in remote corners of the world. Immersing oneself in the raw power and beauty of Mount Yasur's eruptive activity is an experience that leaves an indelible mark on the senses and sparks a deep appreciation for the natural forces that continue to shape our world.

In conclusion, Mount Yasur stands as the island's sentinel—a guardian of fire and spectacle. A visit to this mesmerizing volcano on the remote island of Tanna offers an opportunity to witness its captivating lava display, embrace the power of nature, and gain a deeper understanding of the remarkable geological forces that shape our planet. It is a journey that stirs the soul, sparks wonder, and leaves an everlasting impression of Mount Yasur's mesmerizing presence in the South Pacific.

The Forgotten Volcano: Mount Tambora's Eruption That Shook the World

In the remote Indonesian archipelago, a forgotten giant lies dormant, its name etched in history as the site of one of the most catastrophic volcanic eruptions in recorded memory. Mount Tambora, a towering stratovolcano on the island of Sumbawa, once unleashed a fury that reverberated across the globe. This chapter delves into the geological significance of Mount Tambora, unravels the story of its earth-shattering eruption, and explores the remarkable experiences that await those who seek to remember the forgotten volcano.

Mount Tambora stands at an impressive elevation of 2,722 meters (8,930 feet) above sea level, dominating the landscape of Sumbawa Island in the Indonesian province of West Nusa Tenggara. It is part of the volcanic arc known as the Sunda Arc, which stretches across the islands of Sumatra, Java, Bali, and beyond.

Geologically, Mount Tambora is a stratovolcano, formed by layers of volcanic ash, lava, and debris accumulated over thousands of years. The volcano's formation is intricately linked to the subduction zone where the Australian plate plunges beneath the Sunda plate.

This subduction process fuels the volcanic activity that characterizes the region.

The most infamous event associated with Mount Tambora occurred in 1815 when it unleashed a catastrophic eruption. The eruption began on April 5 and reached its peak intensity on April 10, devastating the surrounding region and leaving a lasting impact on the world. The explosion was so powerful that it was heard over a thousand miles away, and the volcanic ash cloud reached heights of up to 43 kilometers (27 miles) into the atmosphere.

The eruption of Mount Tambora in 1815 is considered the most powerful volcanic eruption in recorded history. The blast resulted in the collapse of the volcano's summit, leaving behind a vast caldera, now known as the Tambora Caldera. The eruption unleashed a deadly combination of pyroclastic flows, volcanic ash, and tsunamis that claimed the lives of tens of thousands of people, devastated local communities, and had far-reaching global consequences.

The environmental impact of the Tambora eruption was significant. The massive amount of volcanic ash and gases ejected into the atmosphere caused a substantial reduction in global temperatures, leading to a phenomenon known as the "Year Without a Summer" in 1816. Crops failed, livestock perished, and famine struck regions as far as Europe and North America. The eruption's effects on global climate patterns were observed for several years, highlighting the profound influence volcanic eruptions can have on Earth's climate system.

Today, visiting Mount Tambora offers a chance to connect with the volcano's dramatic history and witness the remarkable landscapes it has shaped. The Tambora Caldera, with its immense size and unique geological features, serves as a testament to the power of the 1815 eruption. Exploring the caldera reveals a stunning crater lake, lush forests, and remnants of past volcanic activity.

Access to Mount Tambora and the Tambora Caldera is possible through guided tours and treks, typically organized from nearby towns such as Dompu or Bima. These tours provide opportunities to hike through breathtaking landscapes, learn about the volcano's history, and gain insights into the local culture and traditions.

When planning a visit to Mount Tambora, it is crucial to be prepared for challenging conditions. The trek to the caldera involves steep slopes, rugged terrain, and potentially variable weather. It is recommended to embark on this adventure with experienced guides who can provide necessary safety information, navigate the trail, and ensure a memorable and secure experience.

Respecting the natural environment and local communities is essential during a visit to Mount Tambora. Adhering to park regulations, following designated trails, and avoiding littering are important for preserving the area's ecological integrity and cultural heritage.

In conclusion, Mount Tambora's cataclysmic eruption in 1815 left an indelible mark on the world, forever etching its name in history. Visiting this forgotten volcano offers an opportunity to witness the aftermath of its earth-shaking power, explore the Tambora Caldera's unique landscapes, and reflect on the enduring impact of volcanic activity. It is a journey that transports us back in time, reminding us of nature's incredible force and the profound interplay between volcanic activity and the world we inhabit.

The Phoenix's Nest: The Birth and Evolution of Santorini's Caldera

In the heart of the Aegean Sea, a jewel emerges from the depths, a place where myth and geology intertwine in a mesmerizing tale of birth and evolution. Santorini, with its iconic crescent shape and breathtaking views, is a testament to the powerful forces that shape our planet. This chapter unravels the geological significance of Santorini's Caldera, traces the island's captivating history, and invites you to explore the enchanting landscapes of this phoenix's nest.

Santorini, officially known as Thira, is a volcanic archipelago located in the southern Aegean Sea, part of the Cyclades island group in Greece. Its unique geological formations and vibrant cultural heritage have made it a sought-after destination for travelers from around the world.

The birth of Santorini can be traced back to a series of volcanic events that occurred over thousands of years. The volcanic complex of Santorini is the result of a powerful eruption that took place around 3,600 years ago, known as the Minoan eruption. This cataclysmic event, one of the largest eruptions in recorded history, transformed the island and shaped its present-day appearance.

The Minoan eruption was a complex volcanic event characterized by multiple phases. It began with the eruption of a large volume of magma, causing the collapse of the central part of the island. This collapse formed the caldera, a large volcanic crater filled with seawater, which is the defining feature of Santorini's landscape.

The caldera of Santorini is a sight to behold, with its steep cliffs rising dramatically from the azure waters of the Aegean Sea. It measures approximately 12 by 7 kilometers (7.5 by 4.3 miles) and is one of the largest volcanic calderas in the world. The cliffs surrounding the caldera showcase layers of volcanic ash, pumice, and lava, telling the story of Santorini's volcanic history.

Over the centuries, Santorini has experienced subsequent eruptions and volcanic activity, contributing to the island's evolving landscape. The most recent significant eruption occurred in 1950, known as the Kameni eruption, which formed new volcanic islands within the caldera. These islands, named Palea and Nea Kameni, offer a unique opportunity to witness ongoing volcanic activity, with steam vents, hot springs, and the mesmerizing sight of solidified lava flows.

Visiting Santorini allows you to immerse yourself in the island's captivating beauty and rich cultural heritage. The main towns of Fira, Oia, and Imerovigli perch on the caldera's edge, offering breathtaking panoramic views of the sea-filled crater and the neighboring islands. The sunsets of Santorini, with their vibrant hues reflecting off the caldera's cliffs, are renowned as some of the most picturesque in the world.

Exploring the island's ancient ruins, such as the archaeological site of Akrotiri, provides a glimpse into the Minoan civilization that thrived here before the cataclysmic eruption. The well-preserved ruins of Akrotiri reveal a sophisticated settlement with advanced architecture, vivid frescoes, and evidence of a prosperous society that existed before the volcanic devastation.

To truly appreciate the geological wonders of Santorini, it is worth embarking on a boat tour around the caldera. These tours allow you to witness the sheer magnitude of the volcanic cliffs, sail past the volcanic islands of Palea and Nea Kameni, and even swim in the healing thermal waters of the hot springs.

When visiting Santorini, it is important to be mindful of the fragility of the volcanic landscape and respect the local environment. Stick to designated paths, avoid littering, and be cautious when exploring steep cliffs and volcanic formations.

To avoid the crowds and fully embrace the island's tranquility, consider visiting Santorini during the shoulder seasons of spring or autumn. This allows for a more intimate experience, with fewer tourists and pleasant weather conditions for exploration.

In conclusion, Santorini's Caldera stands as a testament to the geological forces that have shaped the island's remarkable beauty. A visit to this phoenix's nest offers an opportunity to marvel at the birth and evolution of Santorini, witness ongoing volcanic activity, and immerse yourself in the rich cultural heritage of this Aegean gem. It is a journey that invites you to explore the depths of Santorini's ancient past and embrace the allure of its breathtaking landscapes.

The Jewel of the Andes: Discovering the Beauty of Mount Chimborazo

Nestled amidst the majestic peaks of the Andes, a towering gem awaits those who seek the grandeur of nature's creations. Mount Chimborazo, standing tall as the highest peak in Ecuador, captivates the hearts and minds of adventurers and nature enthusiasts alike. This chapter embarks on a journey to explore the geological significance of Mount Chimborazo, uncover the natural wonders it beholds, and invite you to discover the unparalleled beauty of this jewel in the Andes.

Mount Chimborazo, situated in the central part of the Ecuadorian Andes, rises to a staggering height of 6,268 meters (20,564 feet) above sea level. Its majestic presence dominates the surrounding landscape, offering breathtaking vistas and awe-inspiring panoramas. The mountain holds a significant place in local folklore and is deeply intertwined with the cultural fabric of Ecuador.

Geologically, Mount Chimborazo is a stratovolcano, composed of alternating layers of volcanic ash, lava, and other materials accumulated over thousands of years. It is part of the Pacific Ring of Fire, a region characterized by intense tectonic activity and volcanic eruptions. The volcano's formation is attributed to the subduction of the Nazca plate beneath the South American plate, resulting in the creation of the Andes mountain range.

One remarkable feature that sets Mount Chimborazo apart is its distinction as the farthest point on Earth from the planet's center. Due to the Earth's equatorial bulge, the peak of Mount Chimborazo is actually farther from the Earth's center than Mount Everest, despite being lower in elevation. This unique attribute makes Mount Chimborazo a magnet for mountaineers and explorers eager to conquer its summit.

Visiting Mount Chimborazo offers a remarkable opportunity to immerse oneself in the natural wonders of the Andes and witness the breathtaking beauty of the Ecuadorian highlands. The mountain is surrounded by a diverse array of ecosystems, ranging from Andean paramo grasslands to high-altitude wetlands, known as páramos. These environments are home to unique flora and fauna, some of which are endemic to the region.

Ascending Mount Chimborazo is a challenging endeavor that requires physical fitness, mountaineering skills, and proper acclimatization to high altitudes. It is recommended to undertake the ascent with experienced guides who are knowledgeable about the mountain's terrain, weather conditions, and safety protocols. The climbing season typically runs from December to February when weather conditions are more favorable.

For those who prefer not to climb, there are several options for enjoying the beauty of Mount Chimborazo from lower elevations. Scenic drives through the surrounding countryside offer opportunities to appreciate the mountain's grandeur and capture stunning photographs. Additionally, hiking trails in the nearby Chimborazo Wildlife Reserve provide a chance to explore the diverse ecosystems surrounding the mountain and encounter unique plant and animal species.

To make the most of your visit, consider staying in nearby towns such as Riobamba or Ambato, which offer a range of accommodations and serve as gateways to Mount Chimborazo. These towns provide opportunities to learn about the local culture, sample traditional Ecuadorian cuisine, and connect with the warm-hearted people who call this region home.

When exploring Mount Chimborazo and its surroundings, it is important to practice responsible tourism and respect the fragility of the ecosystem. Stick to designated trails, refrain from littering, and be mindful of the unique flora and fauna that thrive in these high-altitude environments. It is also advisable to pack appropriate clothing, including layers for changing weather conditions, and to carry sufficient water and snacks.

In conclusion, Mount Chimborazo stands as a beacon of natural beauty in the Andean landscape. A visit to this jewel of the Andes offers a chance to immerse oneself in the grandeur of towering peaks, discover unique ecosystems, and connect with the rich cultural heritage of Ecuador. Whether conquering its summit or admiring it from lower elevations, Mount Chimborazo's allure leaves an indelible mark on the adventurous souls who venture into its realm.

The Firebird's Domain: Exploring the Intricacies of Mount Nyamuragira

In the heart of the African continent, where the lush jungles of the Democratic Republic of Congo thrive, a fiery realm exists that beckons the adventurous spirit. Mount Nyamuragira, a stratovolcano of immense proportions, unveils the intricacies of volcanic activity and showcases the raw power of the Earth. This chapter delves into the geological significance of Mount Nyamuragira, unravels its captivating story, and invites you to explore the enchanting landscapes of the Firebird's Domain.

Located in the eastern part of the Democratic Republic of Congo, Mount Nyamuragira rises to an elevation of 3,058 meters (10,033 feet) above sea level. It is one of Africa's most active and voluminous volcanoes, drawing the attention of scientists and adventurers alike.

Geologically, Mount Nyamuragira is a shield volcano, characterized by its broad, gently sloping profile. It is part of the East African Rift System, a region known for its intense tectonic activity and volcanic eruptions. Nyamuragira's formation is a result of the diverging tectonic plates that have created the East African Rift, allowing magma to rise to the surface and build the volcano over time.

One of the notable features of Mount Nyamuragira is its frequent and often spectacular eruptions. It has a reputation for producing fast-moving lava flows, which can extend for several kilometers, showcasing the volcano's immense power. These eruptions, characterized by the effusion of fluid lava, contribute to the gradual expansion of the volcano's flanks and shape the surrounding landscape.

Visiting Mount Nyamuragira offers a unique opportunity to witness the dynamic nature of volcanic activity. However, due to its remote location and the ongoing volcanic hazards, access to the volcano is limited and requires careful planning and coordination. It is recommended to join organized expeditions led by experienced guides and scientists who have in-depth knowledge of the area and can ensure safety during the visit.

When exploring Mount Nyamuragira, it is crucial to prioritize safety and follow the guidance of the expedition leaders. Protective gear, including sturdy hiking boots, appropriate clothing, and necessary equipment, should be worn to ensure a safe and comfortable experience. It is also important to be aware of the potential risks associated with volcanic activity, such as the release of toxic gases and the instability of the terrain.

The surrounding region of Mount Nyamuragira is rich in biodiversity and offers opportunities for wildlife viewing and nature exploration. The nearby Virunga National Park, a UNESCO World Heritage site, is home to a diverse range of wildlife, including endangered mountain gorillas, chimpanzees, and elephants. Guided tours and treks in the park provide a chance to encounter these incredible creatures and immerse oneself in the natural wonders of the area.

When planning a visit to Mount Nyamuragira, it is essential to consult with local authorities, such as the Virunga National Park authorities and volcanic observatories, for the latest information on volcanic activity and safety conditions. They can provide valuable guidance and ensure a well-informed and secure experience.

Respecting the local communities and the natural environment is of utmost importance. Visitors should adhere to park regulations, support local initiatives, and be mindful of the ecological sensitivity of the region. Additionally, engaging with local communities and learning about their traditions and cultures can enrich the overall experience and foster a deeper understanding of the area's heritage.

In conclusion, Mount Nyamuragira stands as a testament to the Earth's fiery power and showcases the intricate dance of volcanic activity. Exploring the Firebird's Domain offers a unique glimpse into the dynamic forces that shape our planet and a chance to appreciate the raw beauty of the African continent. Although access to Mount Nyamuragira requires careful planning and adherence to safety guidelines, the rewards are unparalleled, offering a once-in-a-lifetime experience of witnessing the majestic realm of this remarkable volcano.

Epilogue

As we come to the end of our journey through the world of famous volcanoes, we reflect upon the captivating stories, awe-inspiring landscapes, and the sheer power of these natural wonders. Volcanoes are the embodiment of Earth's dynamism and the result of complex geological processes that shape our planet.

Throughout this book, we have explored the fiery depths of the Earth, from the devastating eruption of Mount Vesuvius that buried Pompeii in ash, to the majestic beauty of Mount Fuji towering over the Japanese landscape. We witnessed the cataclysmic eruption of Krakatoa that unleashed a wave of destruction, and marveled at the enduring charm of Mount Kilimanjaro, Africa's highest peak.

We ventured into the explosive rebirth of Mount St. Helens, observed the enigmatic nature of Mount Etna, and unraveled the intriguing saga of Eyjafjallajökull in Iceland. We discovered the mighty lava flows of Mauna Loa in Hawaii and explored the sacred realm of Kīlauea. We witnessed the devastating eruption of Mount Pinatubo and marveled at the majestic presence of Mount Cotopaxi. We journeyed to the incandescent crater of Mount Nyiragongo, the lush beauty of Montserrat, and the intriguing ballet of Mount Agung in Bali. We uncovered the frozen inferno of Mount Erebus in Antarctica and traversed the majestic silence of Mount Teide. We marveled at the explosive show of Popocatépetl in Mexico and reveled in the mysteries of Hekla in Iceland. We explored the wrath of Mount Rainier in the Ring of Fire and discovered the beauty of Santorini's caldera. We witnessed the birth and evolution of Mount Tambora and ventured into the intricacies of Mount Chimborazo. Finally, we delved into the Firebird's Domain, exploring the captivating Mount Nyamuragira in the heart of Africa.

Each volcano has its own unique story, shaped by geological processes, cultural significance, and the impact it has had on the surrounding landscapes and communities. We have witnessed the destructive power of eruptions, the beauty of volcanic landscapes, and the resilience of life that emerges from the ashes. Visiting these volcanoes is not only an opportunity to witness the raw power of nature but also a chance to connect with the Earth's natural wonders and deepen our understanding of the planet we call home.

It is a reminder of our place in the grand scheme of things, and the profound impact that geological processes have on our lives.

As we conclude our exploration of famous volcanoes, let us remember the importance of responsible tourism and the preservation of these remarkable sites. The natural environment surrounding these volcanoes is fragile and must be respected to ensure its conservation for future generations. Let us leave no trace, follow local regulations, and support initiatives that promote sustainable tourism.

May the tales of these volcanoes continue to inspire awe and reverence, reminding us of the extraordinary forces that shape our planet. Let us carry the knowledge and appreciation gained from our journey into our own lives, fostering a deeper connection with nature and a greater understanding of the world we inhabit. And so, we bid farewell to the fiery realms of volcanoes, but their stories and majesty will forever remain etched in our memories, reminding us of the boundless wonders that lie within the Earth's fiery embrace.

Fagradalsfjall, Iceland

Thank you for embarking on this journey through the captivating world of famous volcanoes. It has been an honor to share these incredible stories and the wonders of the Earth's fiery embrace with you. We hope that this book has ignited your curiosity, deepened your appreciation for the power of nature, and inspired a sense of awe for the remarkable landscapes that volcanoes create.

Your support and engagement as a reader mean the world to us. We would greatly appreciate it if you could take a moment to leave a positive review and share your thoughts about the book. Your feedback will not only help us grow as writers but also encourage others to embark on their own exploration of the natural wonders that exist in our world.

By leaving a review, you can help us reach more readers and inspire a greater appreciation for the beauty and significance of volcanoes. We value your opinion and would be grateful for your honest feedback.

Once again, thank you for choosing to read this book. We hope it has left you with a deeper understanding of the power and allure of famous volcanoes. May their stories continue to captivate and inspire you as you embark on your own adventures in the world.

Warmest regards,

Jordan

www.ingramcontent.com/pod-product-compliance
Lightning Source LLC
Chambersburg PA
CBHW041941240526
45473CB00033B/117